THE THEORY OF PARTICLE MATTER FREQUENCIES AND MULTIPLE UNIVERSES

THE THEORY OF PARTICLE MATTER FREQUENCIES AND MULTIPLE UNIVERSES

BY ALASTAIR R AGUTTER

QUOTATION

"If the spirit of your imagination is limited, your journey of discovery will be very short."

~ Alastair R Agutter

THE THEORY OF

PARTICLE MATTER FREQUENCIES AND MULTIPLE

UNIVERSES

BY ALASTAIR R AGUTTER

First Recorded and Published 2nd February 2015.

Printed, Published and Distributed by

Create Space Independent Publishing

An Amazon Group Company

ISBN-10: 1507841035

ISBN-13: 978-1507841037

CONTENTS

QUOTATION

"The measurement of human ability is in an infant state. Was it a requirement for the Wright Brothers to hold PhD's before they could invent and take flight?"

~ Alastair R Agutter

NOTE

Sir Isaac Newton's Natural Law tends to have an uncanny ability to throw up into the human mix of discovery and learning, timely events or information and from such revelations, does and can the human race receive a super-charged boost to move on yet again.

The two greatest set-backs in human evolution, has always derived from greed and vanity! These facts have been documented through the ages from the written word detailing conflict and empires.

It is true by understanding your history and past events can you endeavour to advance and evolve beyond such primitive behaviour and this can only be achieved by knowledge and understanding.

Academia and the Sciences tend to try and isolate themselves from the unknown, be it in a spiritual form or religion. Such thoughts and ideas are not alien or primitive gestures of knowledge, but a direct relevance and understanding of the universe and how a human entity is connected.

Greed and vanity began to originate from early forms of tribal control of regions and eventually through natural law evolved to man-made religion and then political governance that exists today in many parts of the world.

The greed and vanity to believe or influence, comes from an idea or an agenda and so often in societies where there is noble effort, but then can be hijacked or perverted for an individual's cause and emerging from that a movement and ideology. Significant examples through history have been the Roman Empire, Catholicism, Kingdoms, Empires, World One and World War Two that gave birth to Nazism!

Today as the human race endeavours to advance for the greater good as a global society, regional conflicts still continue from outdated and aged doctrines of greed and vanity. This is clearly seen in the Middle East, where Kingdoms and tribes continue to fuel conflict to the masses to retain power and influence for only a few.

Modern political governance is still sadly today influenced and seduced by money, commonly described as corruption. However, Natural Law does prove and teaches us that such continued ideologies born from greed and self-interest is not sustainable! For as the masses retain less, there will always come a time of re-addressing such social injustices and again leading to conflict and other forms of governance and control where in recent history, we have witnessed capitalism being replaced with communism and both of which are flawed.

I believe this work is timely; to expand the mind further and to look beyond greed and vanity driven in society, for all that has come before is as mentioned and historically documented where there is evidence of failure.

Natural Law cannot be determined or controlled by the ignorance and vanity of man. Natural Law will continue along a path of evolution and refinement. It is therefore imperative that human society evolve and learn to co-exist with natural law to advance further than any one could imagine.

Academics and Scientists today, are only really beginning to explore and understand Natural Law, where in the melting pot come's Quantum Physics, Quantum Mechanics and Natural Branching that was well understood by Professor John Nash, when feeding pigeons and seeing a pattern of symmetry.

QUOTATION

"The human condition to ignore or dispel ideas is an admission of fear by oneself and refusing to seek out the phenomena of the unknown, to complete the cycle of understanding through greater knowledge and enlightenment."

~ Alastair R Agutter

INTRODUCTION

There will always be views and opinions expressed to fit into one's own ideas and ideologies and if you have bought into a superficial material society driven by money, your well-being can only equate to an existence.

Many of the great scientists and philosophers over the centuries, or in fact since the beginning of human thought, have been exposed to ridicule and as so often mentioned, as a result of fear for the unknown or change.

Sir Isaac Newton's Natural Law, teaches us that nothing ever stops but advances in search of perfection, continuously seeking refinement. This is evident with every living entity on earth and beyond, in the cycles of planetary activity across the cosmos.

The great minds of the past and especially Albert Einstein, had his abundance of critics. But Einstein from enlightenment that has come to others before him, use to say "I am right they are wrong and will go away."

As I said to an acquaintance just recently by understanding natural law, I often take an analytical approach to many activities that are mundane and have applied the delete button on my computer very efficiently, to the utterly stupid and mindless, who continue to email me on search engine optimization offers, when I am an accomplished author on the subject.

I hope from my findings written in this book it will soon become apparent that my theory of Particle Matter Frequencies and Multiple Universes begins to make real sense, for we are surrounded each day with natural and technological events that evolve with similar patterns.

Google is an organization in our modern world that has received a considerable amount of criticism recently, regarding the mundane failed framework in society relating to tax avoidance and evasion. But as an artificial intelligence organization, Google have become one of the most significant entities for the 21st century, especially regarding analytics for human understanding and advancement. Where technology of data is producing analytical results and this incredible feature and function operates in a similar format and pattern to my work in relation to the understanding of particle matter frequencies and multiple universes. For the data gathered, is coming in from many different directions but uniquely operating as a collaboration operation. So in other words, we have a new artificial intelligence entity constantly gathering information and evolving as more data is gathered and with an analytical process of thought similar to how Natural Law functions and evolves.

In a modern day global society, entities such as Google are contributing to a community where results can be produced from the gathering of such data. If you think how humans have communicated over the ages, you can see the evidence for yourself in changes to these forms for gathering and collecting data. This therefore means the human capacity to collect and store data, has a comparable resemblance regarding artificial intelligence of computing in the analytical data gathering process. To have such similar patterns in these two examples and if we drill down and look at all other entities and cycles, there will be found analytical data storage also, but where?

QUOTATION

"The human interpretation of a disability is a primitive departmentalized view of ignorance."

~ Alastair R Agutter

SIR ISAAC NEWTON'S NATURAL LAW

Sir Isaac Newton believed that everything on Earth, in the Universe and beyond is governed by Natural Law and he is right!

Stephen Hawkins when drilling down on his theories came to the conclusion that the big bang was a creation of energy. However any form of energy has to be ignited and by who?

Sadly man made religion has created an environment of taboo's and myths and where many of the events, or facts in scientific terms, have been taken out of factual context or hi-jacked for one's own self-interest and agenda.

Spiritualism in society conjures up witch craft and other mumbo jumbo myths and beliefs that have often been used as a negative in society and so all too often theories or findings have been stifled.

However, as Scientists and Biologists merge more in their scientific discoveries and exploration of the natural world around us today and as more results are gathered, such findings are beginning to unlock many phenomena's and with greater appreciation for the power of Natural Law, so often termed as Mother Nature.

From the tad pole to frog in a life changing evolutionary scenario we understand the transitions and the complex ability and power of Mother Nature, where such a transition involves enzymes and collagens transitioning at the rate of trillionths per second.

So from such further understandings of evolution, should in fact present you with a greater capacity to learn and discover more that at one time all seemed impossible.

Natural Law is forever seeking efficiency and refinement and this becomes more apparent when exploring Quantum Mechanics and Natural Branching where both such subjects are entwined.

From the beginning of the big bang and identifying the catalyst to be energy, all descending events are created by rhythmic patterns and events that Isaac Newton described as Natural Law.

Such understanding by this great Scientist has led the way for others to further explore the wonders of possibility and where again this demonstrates how Natural Law functions in a synergy form. So from such notes, writings and teachings have provided a data bank for human analytical deciphering.

The events through human history are a result of Natural Law and where man has failed to understand these powers and endeavoured to try and defy them have come at a significant cost.

Today's challenge in society derives from Corporate Industrialization and where now more than one third of all life specie entities have become extinct. Such arrogance and defiance will be countered by Natural Law and these events we witness today in the form of climate change and more prevalent resistance disease forms that threaten the human races existence.

If we looked at our planet like a car engine and removed one third of the parts and components, we know it is a scientific fact the engine would not work.

All specie and life forms on Earth are there in unison to provide an environment for all entities to survive and failing to respect such Natural Law is not a sustainable solution for all is connected and related.

QUOTATION

"An informed world is not only for today, but a legacy to shape a better world and future for tomorrow."

~ Alastair R Agutter

ALBERT EINSTEIN'S

MOTION, SPACE, TIME AND OTHER

It is most probably undeniable that the greatest Scientist in the 20th Century was Albert Einstein. Some jokingly say he was "one of them" referring in an endearing way to Alien entities.

But I believe Albert Einstein from his works and writings experienced like many notable persons in history, where they were able to tune into other frequencies to gain enlightenment and answers. So often in history there has emerged a figure at the right time and in the right place and this can be referred back to where Natural Law plays its part in the evolutionary game for coincidence is a speculative proposition.

Once Albert Einstein set the world of Science alight with "The Theory of Relativity" and from that achievement enabled his Scientific and Mathematical credentials to be impeccable and this I believe is why he was tasked with trying to decipher the Natural Law phenomena, so often described as "The unknown or the other" by the science community, as they try to grapple with understanding how Quantum Mechanics functions and works, but in a mathematical calculated sense.

The other, is more commonly known today as Dark Energy, the stuff that exists all around us and between planets and galaxies. Einstein worked for 20 years on trying to create a mathematical calculation on the other and to no avail and to the point of torment. But it required a Scientist of Einstein's stature to embark on such a project, to deliver a conclusion that would be respected in the Science community.

Quantum Entanglement and Quantum Mechanics, in the form of measurement and understanding, is so far to date beyond the reach of understanding, for there are so many facets. However, such phenomena do give evidence of symmetry and a force of Natural Law with paths.

To a physicist, or mathematician, who seeks to gain a formula, or equation for understanding, can cause great frustration and this was the case for Einstein.

In the end, Albert Einstein conceded by saying "God has a sense of humour" regarding the other!

In recent year's fractal maths have also become a prominent force for understanding. For fractal maths can clearly demonstrate by using a linear line of measurement in relation to fractal maths the equations and total calculations can vary considerably.

Fractal maths can and does play a significant part in quantum mechanics even when relating to photons for packet data transfer. Where delivery, speed and times, can vary considerably, throwing up a whole new number of questions, as we further explore the sciences.

Einstein delivered to the World a further understanding of forces, especially regarding space and gravitational events between planets. The folding of space, with the probability of worm holes and the variation of time begins to make absolute sense.

When you encapsulate every aspect and dimension from energy, matter, time, motion and space, you reach a calculated conclusion of understanding that all these events equate to data and the process is analytical that we describe as Natural Law.

So then you have to ask yourself, where is this data and where does this data exist? What part does it play in the grand scheme of things, the big soup as I call it, in scientific and mathematical terms and especially as we look at photons in this form of electromagnetic energies that carries Data?

Perhaps the following answers all these questions!

QUOTATION

"Enslavement can come in many guises and the denial of knowledge is one such tragic example inflicted on humanity."

~ Alastair R Agutter

THE THEORY OF PARTICLE MATTER FREQUENCIES AND MULTIPLE UNIVERSES

My journey for endeavouring to understand the grand scheme of things began as a small boy of around 5 years in 1963 as my Father began to teach me the constellations in the sky where we lived.

So my journey in search of understanding and discovery, asking those searching questions has continued now for over 50 years of my life, but more importantly with an open mind.

Throughout history there have been many telling events and moments where individuals have played an important part in the human evolutionary story. Such people and events I no longer believe to be ones of coincidence but enlightenment periods to help humanity on its journey of learning and discovery.

I believe my findings and conclusions are accurate regarding my theory of particle matter frequencies and multiple universes. However in saying that, I have experienced many phenomenal events to substantiate my findings and understanding and I am sure many other predecessors have also experienced in their life-times similar events, when seeking answers to the big questions.

I am now confident to say that from particle matter, can be formed many universes and these can co-exist at the same time in different frequencies. This in some respects relates to Neils Bohr the Physicist where he proved the moon as one example, can only exist when you actually look at it in the grand scheme of human consciousness, relating to the understanding of atomic structure and quantum theory.

We are all familiar with frequencies in many parts of our lives in relation to radios and televisions to mention just two. We can watch for example the BBC One channel, but know if we tune into another frequency we can change and watch BBC Two. However, by switching these channels and frequencies does not mean to say that BBC One channel no longer exists, for it does and continues with its programming. This is the same principles of how you can structure multiple universes by altering the frequencies of particle matter.

On Earth we have many entities and life forms that have different dynamics in vision, as just another example. Again these are strong examples of how particle matter frequencies and genetic configuration can be altered.

I remember speaking with my Daughter last year. When I began to explain the enormity and potential of a human entity that has evolved over 13.8 billion years and how as a human being, you can appear in many forms. For I explained to my Daughter, how she is seen by me, is different to how she is seen by her pet dog Stanley, or how she is seen by a Bee!

It is now also becoming more understood from Quantum Mechanics that smell derives from atoms tuned and configured with specific additional elements that vibrate in other words the senses of smell deriving from harmonics.

These different visions and smells all relate to frequencies regarding particle matter structure and so you then have to ask yourself why would there not be multiple universes? For it would make absolute sense to have multiple universes, as we see evidence all around us on Earth of Natural Branching in the way of refinement and efficiency and this leads back to Natural Law.

Scientists today are beginning to understand the many facets in Natural Law, Natural Branching, Quantum Mechanics and the subject area of Quantum Entanglement.

By understanding the complexities of the above today in computing, Scientists are now exploring far greater dimensional concepts and by strange coincidence, this equates to all we have discussed so far that equates to data, where folk in the science community are now exploring tesseract, 4 dimensional entity concepts of multifaceted storage.

Because everything in the Universe(s) drills down to particle matter in microscopic form and many particle matter forms such as protons and neutrinos can travel through entities. This demonstrates that multiple universes can co-exist in the same space.

We have heard of many strange events throughout history and read of countless experiences and phenomena's that remain unanswered. Perhaps however this is as a result of how the rules of society have been shaped.

A recent programme I watched demonstrated how Natural Law operates and cannot be altered. This related to the rules of tennis as a game between two players and watching the participant's function regarding the returning of a ball to serving in a game. Amidst all these activities and within the rules of the game exists energy and Natural Law. For regardless of the rules of the game we know how a ball will behave in flight and the energy applied to propel the ball.

If we therefore look at society and its rules, they could in fact hinder potential of other possibilities. This then leads us onto understanding the human condition and the rules society has set. Where the environment mostly consists of labour activity to earn money and this in turn pays for food and other needs within the rules of the society game. From such patterns of existence there is no need for the human condition to seek out and explore further, unless they are involved in shaping a different society with different rules for the betterment of all.

So by exploring greater possibility means greater thought and creativity to achieve such goals, inventions or ideals.

This then leads onto the human condition and wiring of the thought process in relation to the human brain.

It is a known fact that half if not two thirds of the brain is not used and the question we should all be asking is why?

Perhaps like a battery full of compartments regarding energy cells. The reason not all is used, is for the simple fact that the performance (human condition) in the rules of the game regarding society, does not require it based on the everyday existence of most human beings.

The human brain is a phenomenal neuron energy force with the ability to store and analyse data. This analysing is done every day in many people's lives regarding what we describe as judgement. In short decision making when presented with scenarios that require an outcome!

However, amidst all of these logical and calculated thoughts comes other characteristics regarding the human condition, known as gut instinct or intuitive thoughts and this then leads back to the human brain condition being an organic analytical processor and storage device of data.

But where does the gut feeling or intuitive instincts come from? These thoughts are often compartmentalized by society as earlier forms of instincts that existed in the gene code when humans were hunter gathers.

However, this is not true for these instincts are paramount to the continuation of the human species and especially when it comes to reproduction and finding suitable and compatible mates.

The emotion of real love is not an experience that can be dissembled down to an analytical process only. For such emotions in the acts of love would merely feel and appear to be no different to making a decision on the choice of takeaway at a fast food restaurant, or ordering a coffee otherwise.

The cycles of reproduction therefore tune into other frequencies of the brain to create such overwhelming emotions in addition to the logical and analytical processes surrounding the motions and functions of reproduction.

Very often events in the way of dreams, or fleeting moments of vision regarding an event, can be departmentalised as an imagination event. This is how the rules of the game regarding society interprets an event, or condition that requires an answer, but one that is never answered!

However, such events, experiences and visions are often as a result of where other parts of the mind make brief contact in the human mundane processing of repetitive information in relation to a framework known as society.

Many in society (the game) as athletes, or professionals of a trade, or art are often focused in a different way to achieve their goals. Athletes often call it "in the zone" and this is where other parts of the brain function to enable these achievements of more remarkable feats. In fact it relates to wiring in the brain, or should we call it connecting to other frequencies!

By training the brain to call upon these additional powers of energy enables the athlete, creator, or artist to deliver the results they seek to attain.

So we can see that the human condition can aspire to greater things by the training of the mind and utilizing these parts of the brain.

They say many anglers are philosophers as they have time to think. But as a participant of the pastime, anglers learn to tune into frequency examples again in the way of currents, wind lanes and other activities the trained eye can only see. To the bystander passing bye, they would never notice the events taking place, for they have not utilized other parts of their mind to tune into certain frequencies to be able to see such events.

This then leads us onto looking through and beyond the canvas!

The examples given above demonstrates that it is a known fact that the human condition of the brain can be trained and wired to look in greater detail at a subject to attain greater ability.

By continuously training the mind can you begin to look through the canvas as I call it! In a spiritual sense it is known as the third eye. This relates to the focusing of the mind where a left eye is one, the right eye is two and both eyes used representing the third eye. By continuously focusing with an open mind can you begin to see other energies and elements you are oblivious too in your daily lives. In daylight you can

22

begin to see countless millions of photons and energies dancing around you and also other elements, including virus entities and more.

One of the most offensive remarks I find hard to accept is the saying "you only live once and have only one life" and such comments should be consigned to the utterly stupid. For it is a fact that human evolution thus far to date has derived from over 13.8 billion years of evolution and if one is to believe after such engineering, your journey spans a mere few decades, it demonstrates clearly how absurd such a comment really is!

So what does this tell us regarding the canvas we see and can see?

It demonstrates that we in fact live in a sea of energy on Earth, in the Galaxy, throughout the Universe and other Universes. However, humans cannot see such phenomena as they have not trained the power of the human mind to tune into these frequencies. Similar to as we discussed earlier, when we watch and switch to different television channels, for others do exist even though we are not watching them.

So what is this sea of energy and what does it mean to us?

Well within this sea of energy we do have matter and space, but all matter are particles vibrating at different frequencies to deliver through photons an interpretation of vision to what exists.

But then this leads us back to the big bang question and where Stephen Hawkins believes once you deduct space and matter, you are only left with energy and a positive energy and a negative energy and when you deduct one against the other, the sum of all equals zero. This relates then to the big bang being created by energy only.

However, regarding this event Stephen Hawkins is wrong. Energy cannot function unless there is another element and that is data! For it is data, that is the intelligent force entity that is the only element to give instruction to execute the process. For the big bang to happen!

There had to be intelligence in the form of data to carry out the instruction. I describe it like a mixing bowl where you place all your ingredients in to make a cake. The spoon is the energy force to mix the ingredients. But only by the instruction of intelligence, can this process begin and function by the spoon, then stirring together the ingredients.

So how can we begin to comprehend and understand this sea of energy phenomena?

The best way to visualize such an environment is by looking at the World Wide Web. Where there are countless of millions of computers and servers sharing data across the World Wide Web and all being processed and yet at the same time, also using different operating systems and programs to perform these functions.

Google for their part shows us how amidst all of this technology there can also be an analytical process of data gathering to refine results to seek answers and solutions. The same process is in operation on the World Wide Web as what we call Natural Law regarding Earth, and the Cosmos, where every entity is functioning within the rules of Natural Law, but at the same time evolving in the way of development and refinement.

So what can we see through the canvas if the mind and human condition is wired as so?

Albert Einstein's closest friends and acquaintances knew or believed that Einstein could visualize the whole of this Universe when he tuned into these frequencies. But I am sure he was able to see many others. In a spiritual sense it is known as Astral Plaining and where it is possible to see the energy forces, fields and data around the entities.

Once the human condition becomes wired to such understanding and where the minds energy forces begin to ignite the visualization and including other senses that become stronger and greater, this then leads onto making contact with other frequencies and here we begin to explore Einstein's theories regarding time and portals but also telekinesis.

However, regarding Albert Einstein, I believe his knowledge of enlightenment enabled him to tune in and join various frequencies that he described as portals in a scientific term, to protect his integrity from the cynical masses. Encompassed with this belief also relates to time! For you can tune into these frequencies where contact can be made in the past and future, this is not as strange as it sounds.

Albert Einstein wrote and warned the US President, that Nazi Germany were aiming and planning to build a Nuclear Bomb. These warnings were far earlier than the Manhattan Project ever began. It could easily be surmised that Einstein's train of thought were along those lines at the time, but I believe this not to be so, for I am convinced he was applying his thoughts more to outcomes and events, as he further grappled and came to terms with Natural Law.

To try and understand how past and future regarding time can best be described regarding this sea of data we are in, is to imagine that your life's journey is on film. All is being recorded in a sophisticated way regarding your every thought and action.

This is not too outlandish to comprehend for even reading this book you are utilizing the human condition of consciousness to power energy forces and particle atoms for you to see and read such a document. Even your actions now, from flicking, or moving onto the next page of this book, or having a coffee while you read all relates to vibrating particle matter. So even if you read a passage out loud, or have a conversation with someone around you, these are all particle energies that have been configured in a way for this process to function and so your thoughts and actions are being data collected and recorded. Once generated in particle matter form these events cannot just simply disappear, it is a physical impossibility!

So then we need to think of the life's journey of a person, both in film and data form. Let's say we have a film of someone's life on a roll. If we unrolled it, we would see countless millions of pictures and slides recording events and in those events, are others in the form of relatives and acquaintances and they too have their own individual recorded film of their lives. However, from one film to another, they over lap and found is a string of data connected. For an event shared by you and a relative is also on their film roll mirroring the event. This then leads to another interesting consideration where scientists believe and know particle energies have to have a mirrored particle partner so you have two particle energy forms and one being a positive and one being a negative for them to interact and in fact they can be in different places in the Universe(s) or as like the films of you and a relative in two different places at once but both sharing the same event and slide in your film recorded data.

The more you begin to expand this understanding, can you see how all is related for if you move through the slides on the roll of film you will then come across loved ones now past on, that is described in human terms as death, but before your eyes can you see a record of data from such events and as mentioned before it is a physically impossibility for that data collection of events in particle matter form to disappear.

There are many wives tales and mythical stories of the past, where sayings have derived and one none more so than "what goes around comes around" and if you relate to your filmed life history recorded within that film comes as mentioned relatives, friends, acquaintances and events all recorded. Where from those meetings and events, they become joined with others filmed life stories recorded and if we related that in similar terms to the World Wide Web today, we would describe them as data strings!

The more you look into your filmed life story of data recorded, you will find amass of data and all linked to others in the form of data strings until you begin to comprehend that all is connected, even the filmed life recorded of an animal or pet and amongst their filmed data story are other pets, creatures and animals.

The sea of data when wired correctly can be seen and acts like a fine mesh shimmering in the daylight where data is moving up and down constantly. Whereas most can only see the obvious before them, as they have sadly consigned their lives to acceptance and no longer expand the human condition to learn and explore more!

Through an open mind and by tuning in can you connect to these data strings, as they are joined scientifically from particle matter and as an analytical maze of data is a living organic computing entity you can receive data from that is past, present and future!

How can it be the future?

This is where it gets further interesting and leads onto multiple universes, or should I say frequencies, or even storage.

If we look at the World Wide Web, every computer connection has an IP address to identify the computer. On each computer is a hard drive and where data is stored. Now today we see cloud computing most aptly described in the ether as they say stored on large web file servers and where access is gained through communication in the form of handshakes where one IP address connects to another to exchange data.

Well, if we look and understand Natural Law, it has the sole mission of evolution and refinement and so it makes sense to have multiple Universes in the same environments as an efficient way of data collection and storage.

We see such examples all around us from colonies of plants all grouped together to human housing development in the form of skyscrapers. Now when tuning into different frequencies, it is apparent and obvious that the future, past and present are of logical consequence as a result of analytics.

The future does not yet have to exist to receive warnings, or insight into the future regarding Earth and this one Universe you currently exist within. For the organic energy data that we call the big soup or grand scheme is constantly processing data for efficiency that we call Natural Law and so doing, able to produce accurate results when seeking a data request for information deriving from an analytical process be it the future, present, or past.

So from understanding the power of tuning into these frequencies, this then possibly answers many questions regarding known creators, inventors and discoverers of the past. Where subconsciously, or intentionally, they have tuned into these frequencies to gain enlightenment and insight to solve the creation, or event problem that they were confronted with.

It is a fact that many scientific creations today, first started out as science fiction and science fiction is only created from the imagination of the mind. Therefore it should be asked, did the imagination use the power cells and energies within the mind to connect to these frequencies to gain an analytical answer.

When you can wire yourself to tune into these frequencies to receive data very often it can be analytical, where you have to decipher the data. However, data can be transferred to you visually, by smell, or even by harmonics.

Over the ages there have been moments in history where individuals have been blessed to tune into frequencies to gain enlightenment. So these events are nothing new from Jesus Christ, to Mohammed, to Washington, Benjamin Franklin, Abraham Lincoln, Albert Einstein and many others.

However, sadly of such events and learning as we see today, honourable sincere enlightenment to advance humanity can be hi-jacked instead and in pursuit of personal greed and power. The afore mentioned persons in critical times of the human story were given wisdom to help all, but sadly very often others around them took such knowledge for their own dastardly deeds.

Like all data sometimes you can get configuration problems and this I believe to be the case when sometimes folk see a vision of entities from another time, where for some reason frequencies have clashed, or collided. Just as you can on a hard drive of a computer, where there becomes a cluster, or configuration problem of a few corrupt files.

Or more recently, finding the answers to another question often posed and asked regarding ancient monuments beyond the ability of creation in a specific time in history. Or the vision of alien entities commonly described as UFO.

Perhaps these answers are not so difficult to resolve now if we understand it is possible to tune into other frequencies. For an advanced civilization or in fact our ancestors of the future would no doubt then have the ability to isolate a vessel with particle matter forces to jump through to other frequencies, hence the visions of UFO's and the continued monitoring of our troubled world.

The answer to the ancient buildings such as the Pyramids came to me, when I saw the NASA team down on earth send an instruction to a computer in space and then in the space centre, a 3D printer produced a component object required.

It made sense to me and is not impossible if you have the co-ordinates and the correct frequency like an IP address to fire off instructions to assemble such objects by advanced civilizations.

ACKNOWLEDMENTS

"Sir Isaac Newton, Albert Einstein and Neils Bohr, NASA and Google."

www.ingramcontent.com/pod-product-compliance
Lightning Source LLC
Chambersburg PA
CBHW070757180526
45168CB00004B/1650

*9 7 8 1 5 0 7 8 4 1 0 3 7 *